Training Manual

The Intensive Airless Spraying Course

ASSURED

Paint tech training academy
www.painttechtrainingacademy.co.uk
23/07/2020
6 X 9

Aim

Our aim is to give decorators the skills and knowledge needed to allow them to successfully spray walls, ceilings and woodwork using an airless sprayer.

Course objectives

1. What are the main advantages to spraying?

2. Understanding the spray machine and all its parts.

3. Choosing the right tip, filter and pressure.

4. Adding accessories for maximum gain.

5. Working through common objections.

6. Masking products and systems.

7. The correct safety and PPE.

8. Setting up the sprayer and cleaning correctly.

9. Spraying technique over various surfaces.

What are the main advantages to spraying?

Speed
Quality of finish
Less staff
Health
More fun

Speed

We don't just mean it's a bit faster it's a lot faster. We could quote lots of examples but the easiest to use is a panel door. Let's assume its bare wood and it needs 4 coats. 1 primer, 1 undercoat and 2 glosses.

To do this by brush would take over just over an hour. That's assuming 15 - 20 mins per coat.

To spray one side of a door would take 30 seconds. Yes, that's half a minute. Now I know what you're thinking, there is loads of masking to do. However, it would not take us much more than 2 or 3 minutes to mask a door. We would take the handle off (we would do that anyway even if we were brushing it), mask the hinges and put some paper under the door. We would do this anyway too.

To sum up, let's assume that with masking, a door would take 5 mins for 1 coat. There is no masking with the second coat, it's already done. So, 4 coats would take 6 mins 30 seconds. Let's say 10 minutes.

Brush – 60 minutes
Spray – 10 minutes

Imagine if you were doing a new extension and you had 20 doors, both sides!!

Quality of finish

One of the main selling points for the customer is not the speed, it's the finish. This is a big advantage of spraying. We are living in an age of the fussy customer. They are paying us a lot of money to decorate their property and they want perfection. Most decorators we know can give close to perfection with a brush. However even the best decorators when using water based paints cannot match a sprayed finish.

We know what you're thinking: - "Hang on a minute, why did you slip in water based paints? I always use oil, water based is rubbish" this has been the view of many decorators over the years. Unfortunately, many oil paints yellow quite quickly due to modern formulations, this means you cannot really use them on an interior decoration job and get lasting results when using white.

Enter the sprayer. You can spray acrylic paints and get a perfect finish. The finish is so good that customers will talk about it weeks after you have gone. How good is that? Your woodwork is recommending your services to your previous customer's friends and family for you.

It's not just about the woodwork. Ceilings look great sprayed. The emulsion flows out and you get a perfect finish. Once you get used to this finish, a roller finished ceiling looks terrible.

For those 2 reasons alone it's worth biting the bullet, getting yourself a little sprayer and having a go. You will not look back.

Less staff

Looking at the above example with the doors ask yourself how many staff you would need to paint the same number of doors in the same time. If you are producing work 4 times faster, then you need 4 times less decorators to produce the same output. If you want to expand your business but are struggling to find good staff, then spraying can be the answer.

Health

Rolling ceilings or painting skirtings on your knees can take its toll over time. Because spraying is physically easier to do you will find that you will produce the work with less effort and less exertion on the body.

Fun

A lot of decorators once they have done the course contact us to tell us about an additional benefit that we had never mentioned. It's fun. They are finding that they are enjoying decorating again. Better money, no more rushing about, a great finish and fun to do as well.

Understanding the spray machine and all its parts

Airless sprayer

First of all, what is an airless sprayer? This is a sprayer that does not use air at all, hence "airless".

The system is basically a pump that pumps the paint down a hose at high pressure to the gun. Because the paint is under so much pressure when it reaches the gun, it atomises.

Think of a pressure washer that you use to wash your car, it's basically the same principle.

The advantage of airless is that it is very fast. The disadvantage is that it can be harder to control and because of the high pressure (typically 2000 psi) it can be dangerous because of the risk of injection hazard.

Treat the equipment with a lot of respect. Lock the trigger when not using the gun.

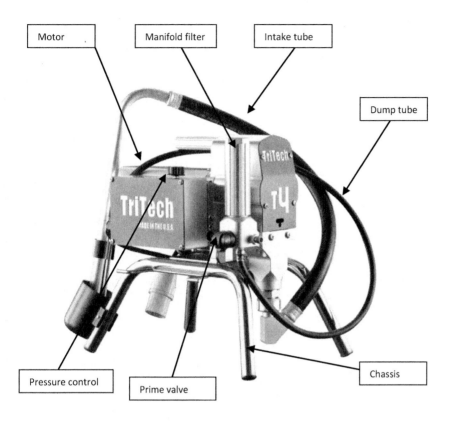

Here are the parts of an airless sprayer.

An airless sprayer is made up of a number of parts.

These are:-

The pump

This is a pump (the wet end) driven by a motor and it's the part that pumps the paint from the tin to the gun.

The hose

A standard hose is 15 metres long and this is the size of hose you will typically get when you buy a new sprayer. However, this is not always a convenient size. Sometimes you will need a longer hose, in this case you can connect 2 hoses together to make a 30m hose.

Sometimes you need a shorter hose, when you are just working in one room for example, you can buy shorter hoses, 7.5m.

The gun

The gun is the business end of the system and this holds the tip and filter and delivers the paint to the surface. Most guns are pretty standard, and you can mix and match different brands of gun to suit yourself.

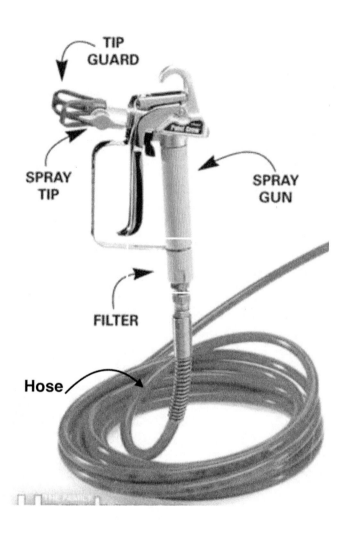

TIP
GUARD

SPRAY
TIP

SPRAY
GUN

FILTER

Hose

Airless spray guns

Four finger
trigger

Adjustable

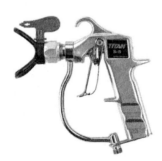

Spray plaster

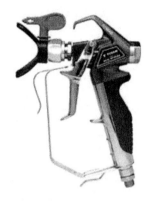

Two finger trigger

HEA gun

Inline gun

There are some minor differences that you need to be aware of.

Graco tip guards and tips are different to other manufacturers so if you have a Graco gun and some Titan tips then these will not fit a Graco tip guard. All is not lost though because you can buy a Titan tip guard (or any other brand of tip guard) and this will screw onto your Graco gun allowing you to use any brand of tip.

Graco filters are also different to any other brand. This is no big deal but it's worth being aware that you need a special filter for Graco guns, and the cheap standard filters that are available to fit every other brand of gun will not fit Graco.

The older Wagner guns have their own thread size for the tip guard. This size is $11/16^{th}$ and you may see this on some websites when buying extension poles. The standard thread of $7/8^{th}$ is what most guns are - and even the new Wagner guns are this size.

Changing the seal in the tip guard

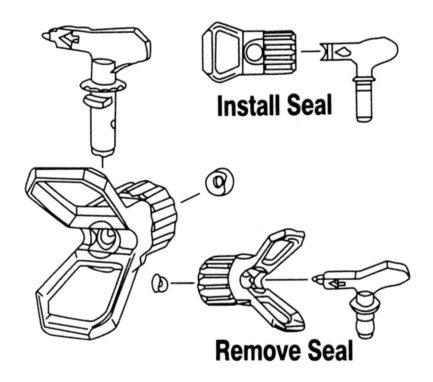

Install Seal

Remove Seal

All guns have 2 safety features that MUST be used, these are: -

1. A tip guard - this is to make it more difficult to get an injection injury and it keep the tip away from your skin.

2. The trigger lock - this must be used whenever you are not spraying. The locks on different guns vary in the way that they work but they all lock the trigger so that it cannot be accidentally pulled.

Choosing the right tip, filter and pressure

The tip

The tip is the key to the whole airless process. Without the tip you just have a hose pipe. The tip that you get with your new sprayer will be a 517. This is a good size to use if you are spraying emulsion on a normal sized wall or ceiling.

You need to understand the number, it's quite easy. 517 is in 2 parts:-

5 – This is the fan width. To find the fan width you double the number. 10" fan in this case.

17 – This is the orifice size, 17 thousandths of an inch in this case or 0.017" The bigger the hole the faster the paint will come out. Also, different paints will need a different size. Usually the paint instructions will tell you which size that you need.

Typical tips we use are: -

517 – Emulsion on ceilings

515 – Emulsion on walls

310FF – This is a fine finish tip and is for woodwork. This would be a 6" fan (3 doubled) with a 10 thou orifice. (Fairly slow).

You can go bigger on ceilings once you get more confident, maybe a 619.

Tips for woodwork can be chosen from a range depending on the size of the surface – 110, 210, 310, 410 & 510.

The filter

There are usually 3 filters on an airless system.

1. The rock stopper (at the end of the intake pipe)
2. The manifold filter (not on smaller sprayers)
3. The inline gun filter (the most important)

The inline gun filter is important. Without the filter your tip will keep blocking with bits from the paint.

Filters come in different sizes:

Red is extra fine or 180 mesh (this is the size of the mesh, the bigger the number the finer it is.)

Yellow is fine or 100 mesh.

White is medium or 50 mesh. You probably got one of these with your new gun.

Green is coarse or 30 mesh.

We generally only use white and yellow. The mesh sizes match the tip size. For example, a white filter is 50 mesh and allows particles through that are smaller than 17 thou. Anything bigger is filtered out. You would use a white filter with a 517 tip.

If you left the white filter in the gun and changed the tip to say a 515 tip then this could block the tip because a 16 thou particle would go through the filter but not through the tip.

A yellow filter would allow particles smaller than 15 thou so this would not block the 515 tip. It's important to use the right filter for the tip.

The smaller the tip, the finer the filter.

Adding accessories for maximum gain

Once you have bought your sprayer you will need some accessories. Of course, you do not have to buy these, you will get all you need in the box when you get a new sprayer however these few items are worth getting to make life easier.

Transformer

You can buy a 240V sprayer however if you are going to be working on site (likely) you will need a 110V sprayer. If you have one of these, you will need a transformer.

Whip hose

This is a short but very flexible hose that goes between your main hose and the gun. High pressure hose is very stiff and can make your arm ache with continuous use.

A whip hose is very flexible and makes moving the gun much easier. You should be able to get one for about 20 - 30 pounds.

Extension bars

These are the spray equivalent of a roller pole. They allow more reach. We find also that a short one makes spraying walls much easier with less bending and stretching. You can get one of these for about 40 pounds. Shorter ones cost a lot less.

Spare tips

We are going to talk about tips in a bit. You will get a standard tip with your sprayer but it's worth having a few different sizes in your toolbox.

Spare filters

Same again with the filters, have some spare ones. These are really cheap, less than a fiver so there is no reason not to get some.

Graco Cleanshot valve – The 'swiss army knife'

One of the advantages of spraying is that you can add an extension pole on the end of your gun and reach high ceilings and walls from the floor. However, you will find that when you add the extension pole then it will start spitting.
This is because when you release the trigger, the valve closes at the gun but there is still pressure in the pole, so it spits. The solution to this is the Cleanshot valve which screws onto the end of the pole and has a valve which closes when you release the trigger.

These are not cheap, however it's worth thinking about getting one once you get going.

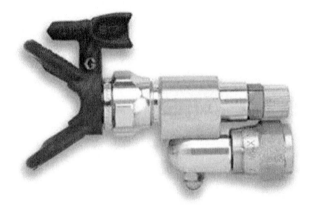

Hoppers

A hopper can be added to most sprayers, these are very useful if you are spraying with smaller quantities of paint or if you are moving the sprayer around a lot. A hopper can make the cleaning process easier too.

Spray rollers

Can be useful when spraying large block or render surfaces, they minimise overspray and make it easier to push the paint into the surface. The paint is pumped to the roller, so you get a constant paint flow.

Working through common objections

Once you hear the advantages of spraying and realise how many decorators in other countries use it to be very productive, then you may wonder why don't we do this more often?

When we talk to decorators, we find that there are a number of well-rehearsed objections. It's worth looking at them now so that you understand where they have come from.

1. You cannot touch up

This is a common one. We understand it to a certain extent. But before we discuss it consider your car. Imagine if you picked it up from the dealer, brand new and shiny but with a roller finish.

"Why have you done this?" You ask the dealer.
"Well the paintwork can get damaged while we make the car, so we do this so it's easy to touch up." Says the dealer.

Really?

We think this issue depends a lot on the type of work that you do, and that you can decide if you want to make it an issue or not.

On new builds, there is often a separate team who "snag". In this case, it may be worth "back rolling" the final coat to allow then to touch up.

New high-end work. If you are the last trade on the job and the customer expects a high standard of finish, then you could give them a flawless spray finish.

2. Masking takes ages

We hear this all the time. We post a video on YouTube of a room being mist coated and then comments come rolling in. "Oh yes" they say "I can see how fast that is" and then add "but I bet it took you all day to mask"

This is such a common misconception that it is very difficult to persuade someone that it is not true. Why is this?

When a decorator decides, they are going to spray they are new at the whole masking thing. They are basically amateur maskers. Of course, it takes them ages, they are not very good at it. They then abandon the whole thing and spend the rest of their lives telling everyone how slow masking is.

If you watch an experienced masker, the one thing you will notice is that they are fast. Check out some of the videos on the academy website or on YouTube. Once you perfect your masking technique then you will fly.

3. There will be loads of overspray

This is a common fear amongst new sprayers. Spraying is messy, and you will get overspray over everything. We understand where this comes from, if you're not careful and your technique is poor then you will get overspray everywhere. There are several things you need to do to prevent this.

    a. Mask up properly. Cover all the areas that you do not want to be sprayed.

    b. Keep the gun the correct distance from the surface. Too far away and you will get overspray.

    c. Use a new tip. Old tips create overspray. Some sprayers use a new tip on every job. While we don't do this, it is worth being aware of how worn your tip is.

    d. Use a shield. On corners, it is easy to spray past the wall and create overspray. If it is critical that you don't do this, on an outside for example then use a shield.

A Shield

4. The equipment can break down.

This is a concern for decorators. It is important that you look after your equipment, use throat seal daily and clean it out properly, this way it will give you hours of trouble free work.

It is worth noting that even though your van can break down no one considers having a cart like they did years ago, to save on the running costs of a van.

5. The sprayer uses more paint

An airless sprayer puts a good coat on the surface. We think that this is an advantage however it can mean that you use more paint.

This can be controlled of course. However, the increase in productivity and the savings in time far outweigh any material costs.

6. The equipment is expensive

This is not a myth; the kit is expensive. However, you need to look at this like any investment. The sprayer is tax deductable, and it's written off over 4 years. So, a £2000 sprayer costs £500 per year.

Then you need to consider the increase in productivity. This is not slight. Even if you're pessimistic and only think that the sprayer will double your productivity then you would make an extra £600 in a week at least.

(Using a basic £120 per day which can vary from region to region and decorator to decorator, use your own day rate for this)

Masking products and systems

Masking is very important. It is in some ways more important than spraying. You do not want to spend hours cleaning off paint where you have not masked properly or touching up where your masking has gone on the wall. **Take care when doing it.**

We use 5 types of masking materials on a regular basis, these are: -

1. Tapes
2. Films
3. Paper
4. Spray glue
5. Floor protection

Tapes

There are standard tapes, precision tapes, low tack and specialist tapes. Precision tapes are for places where you want a good line, tops of skirtings for example. We use standard tape in places where it does not matter as much or if it's just a one or two day job.

Films

We use tesa pre-taped masking film. This is a plastic sheet with masking tape on one edge. This is good for windows, so that you still get light coming in the room and for prefinished doors. You can wrap the doors in the plastic, and they are completely protected.

Paper

Masking paper can be used with a hand masker. The masker combines the paper and tape so that it's easy to mask things. We use this to mask pendant lights, under doors, along the floor etc. Masking paper is fairly cheap, about 3 pounds a roll for 50 metres and 300mm wide. You can get bigger rolls of paper and use them on a masking trolley.

Floor protection

This includes Correx, X-board, hardfloor protection and carpet protector. There are a number of brands for each category and its worth shopping around for the best value.

Hand masker

We use a hand masker for a lot of the masking. The one we use is the 3M, M3000. These are quite expensive at about £50. They can be bought from the 3M website, Amazon or EBay.

There are other makes of hand masker and we have tried them however we don't think any compare to the M3000. The one shown below has a ladder hook which is really handy.

A common question when it comes to spraying is "Which order do I do things?".

There are many answers to this question, and you need to experiment yourself with an approach that suits you.

However here are a few pointers……

Traditionally a decorator will work from the top down. Ceiling, then walls then woodwork. There is a good reason for this, and it is that paint will splash downwards so you will splash the walls when finishing the ceiling, splash the woodwork when you finish the walls etc.

Another reason is that it is easier to cut a wall into the ceiling than the other way around.

When you start spraying, cutting in is replaced by masking. It is easier to mask the walls and spray the ceiling rather than the other way around.

Because of this the following approach works well.

Strategy 1

Finish the woodwork first, then mask it off with paper. Finish the walls then tape and drape the walls and finish the ceiling. This is a good method and it is remarkably quick to tape and drape the walls using tesa easy cover plastic with tape attached.

The only drawback of this method is that if the paint has not fully cured there is a danger of the paint pulling when you de-mask. This is especially true if you are working in a house in January with no heating on. Not really an unusual situation in our game.

Strategy 2

You can first coat the walls with the sprayer, then spray the ceiling to a finish and then spray the woodwork to a finish. Once the woodwork has cured tape the top edge with 25mm tesa yellow. Then cut in and roll the final coat on the walls.

This system has a couple of advantages. It's quick as you're not masking as much, and it means that the walls are easier to touch up without flashing.

Correct safety and PPE

We are always being reminded about the health and safety aspects of everything we do in the workplace and at times it can get tiring. However, when it comes to spraying you need to pay attention.

The two main hazards are: -

1. Breathing in the atomised paint.
2. Injection injury.

Masks

When you are spraying you need to wear a proper mask.
There are a number on the market so it's worth trying
them out to see which you find comfortable. We are
going to recommend 2 masks.

1. The JSP Force 8 mask

   This is a vapour mask and is very comfortable to
   wear.

2. The 3M range of masks.

   These are available at many places, including Amazon.

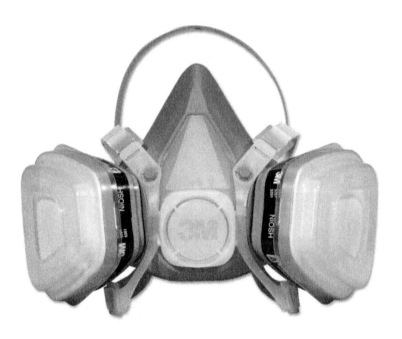

High pressure

You need to learn to respect the fact that an airless sprayer works at very high pressure. Never direct the gun at yourself, your hand or anyone else. Always put the trigger lock on when not spraying and release the pressure in the system when not in use.

Make sure that you check all the hoses and connections before you get going, and if you find any faulty fittings then replace them. Never attempt to repair an airless hose.

Injection injury

This is not a common injury but if it does happen it can be fatal, hence it is important to understand how it can happen, how to prevent it from happening and what to do in the worst case that you get an injection injury.

How it can happen

The paint comes out of the gun at a very high pressure, typically 2000psi. At this pressure if the tip is put against the skin and sprayed this will inject you with paint. Once paint gets into the body in this way it is very difficult to remove without amputation. Fluid leaks from the hose can also cause a similar problem and for this reason you must take good care of your hoses and replace them if they get worn.

How to prevent it

Always use a tip guard on your gun, never mess with the tip guard or tips without first depressurising the sprayer. Always use the trigger lock when you are not spraying. Always check all connections and hoses for leaks before starting to spray. If you have a leak, then depressurise the machine and then tighten up the connection. If you have a split in the hose, then depressurise the machine and replace the hose with a spare one. Always carry spare hoses.

Remember follow a 6-point safety check

1. **Turn the trigger lock on**
2. **Turn the tip halfway**
3. **Turn the pressure control down**
4. **Release the pressure using the prime valve**
5. **Turn off the machine**
6. **Unplug the power**

What to do if you get an injection injury

If you do get an injection injury you will know because it is very painful. You need to go straight to hospital and give the triage nurse the card that comes with your sprayer explaining what injection injury is and what to do about it. Do not hesitate to go to the hospital.

Gloves and glasses

You will want to wear safety glasses if your spraying a ceiling, and gloves while working to keep your hands clean.

Setting up the sprayer and cleaning correctly

This is a quick set up guide for your sprayer. Make sure that the piston has been oiled. You need a few drops every day that you are using it. We usually oil at the start and end of the day.

Attach your hose and gun, plug in your sprayer and prepare your paint. Have a bucket of water on hand.

1. Put the intake hose into the paint. **Turn the pressure down low.** Make sure that you are in prime mode and switch on. The paint will circulate round the pump. Allow this to go on for about 30 seconds. Switch off.

2. If your sprayer is new you will not have any water in the hose. Do not put the tip in the gun yet. Put the end of the gun under water slightly in your water bucket and pull the trigger. Set the sprayer into "spray" mode and switch on. Wait for the paint to appear in the water and then switch off. **Remember trigger first and then switch on.** This prevents the pressure from building up.

3. Put your tip in the gun and do a test spray on a piece of masking paper. **Turn the pressure up until you get a nice solid "letterbox".**

4. You are now ready to spray.

To sum up: -

1. **Put the suction hose in the paint**
2. **Switch to prime and turn on the sprayer**
3. **Once the sprayer is primed switch to spray**
4. **Run water out of the hose until you get paint**
5. **Put the tip into the gun and set the correct pressure and test the spray pattern**

Cleaning the sprayer is the reverse of this process.

1. Take the intake tube out of the paint and put into a bucket of warm, clean water. Put the return tube into the tub of paint, **we are going to empty the paint out of the pump first**. Switch to prime mode and switch on. It will not take long for water to start coming through. Put the return tube in the waste water bucket.

2. We are going to empty the paint from the hose. Remove the tip from the gun. Aim the gun into the paint bucket and pull the trigger. **Remember trigger first and then switch on.** Then switch the pump to spray and switch on. The paint will pump into the paint tub. Once water starts coming through switch to the waste water bucket.

3. Once the water has run through the hose you need to change the bucket for clean water. When you do this also remove the manifold filter for cleaning and the gun filter. Get some clean warm water and clean your filters.

4. Replace the manifold filter on the machine and then run the second bucket of water through. This time leave the machine in prime mode and pull the trigger on the gun. The water will circulate through the pump and the hose at the same time.

The second bucket of clean water should do the trick and clean the pump. The water should run clear.

Spraying technique and spraying pressures

There is a lot to think about when you first start spraying with an airless sprayer. It's very fast so it does not give you much time to think. Here are the 4 things you need to think when you're spraying.

1. Distance

It is important that your gun is the correct distance from the wall. If it is too far away, then you will get loads of overspray and if it's too close you will get runs. The correct distance is about 12" or 30 centimetres.

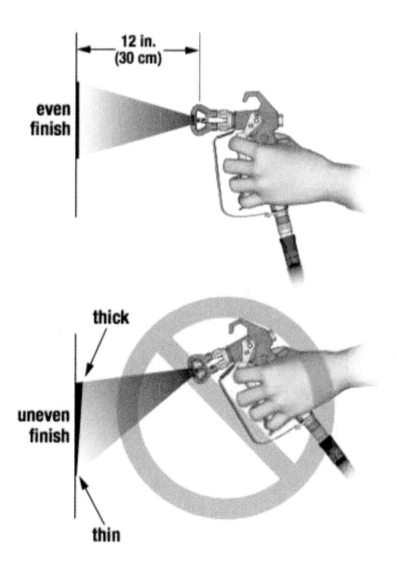

The most common mistake that beginners (and some experienced sprayers) makes is having the gun too far away from the surface. This does get paint onto the wall and feels like you are doing it right however you are putting a lot of paint into the air and this causes overspray and also wastes paint.

2. Don't arch

The gun needs to be parallel to the wall all the time so that the distance remains constant. Unfortunately, your arm naturally forms an arch when you swing it and you don't really know that you are doing it.

Keeping the gun, the same distance from the wall does not feel "natural". Look at your spray pattern on the wall and if you're getting a nice straight line then you're on the right track. This is a very common beginner mistake.

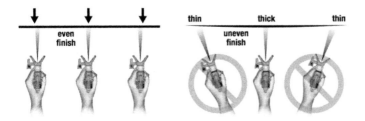

3. Keep moving

When you pull the trigger, the paint comes out of the gun fast. 5 litres a minute in some cases. That's a lot of paint. There is a tendency when learning to hold the gun still, then pull the trigger and then start moving.

Unfortunately, this creates a heavy patch of paint at the start of the stroke which results in runs.

To prevent this, you need to be moving when the trigger is pulled. Of course, you then need to release the trigger before you stop.

Once you get into the trigger on, trigger off routine, it does become second nature. Some people are tempted to keep the trigger pulled and just spray constantly. Unfortunately, when you change direction you pause for a second creating a heavy patch.

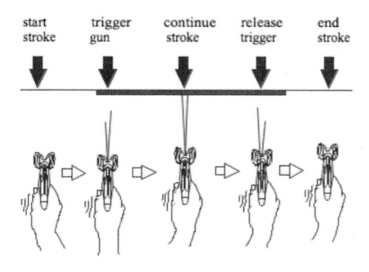

| start stroke | trigger gun | continue stroke | release trigger | end stroke |

4. Overlap by 50%

Each pass needs to be overlapped by 50%. This is easy to do if you aim your gun at the edge of the previous pass. Most people get this pretty quickly.

5. Control your speed

The speed at which you move the gun controls the amount of paint that you put on the wall (other than the tip of course). The faster you move the less paint you put on the wall.

There is a tendency to move the gun slower to get the paint to cover and while you will probably get away with this on a ceiling it can be dangerous on walls or doors as you can end up putting too much on.

These are the 5 basic things to think about when you start, it seems like a lot but you soon get the hang of it. Good surfaces to practice on are new plaster and ceilings. The new plaster is very absorbent and therefore is less likely to run.

What pressure do I spray at?

Typically, you will spray at around 2000 psi. This is the usual quoted pressure that the airless sprayer works at.

You will find though that you will spray at a range of pressures from 800 psi to 3300 psi

The main factors are the size and type of tip and the thickness of the product that you are using.

Standard tips and Low pressure tips.

Standard tips end with an odd number, 517 for example and are used to spray emulsion. Low pressure or fine finish tips end with an even number 310 for example and tend to spray at lower pressures.

A thick material such as spray plaster will need a lot of pressure to atomise it and a thin material like primer will need less pressure to atomise it. There are many tip, product and pressure combinations but I will give you two to give you an idea how the pressure is set.

Normally you would spray emulsion with a 517 tip, if you did not thin the emulsion much then you would need to spray at 2200psi to get atomisation.

If you were spraying woodwork using a thin wood primer and a 310 ultrafinish tip then you could spray at around 800 psi.

Finally

This is not an exhaustive manual and does not cover absolutely everything that you may come across. It is designed to get you started spraying and to act as a quick reference while you work.

For more information you can get the book "Fast and Flawless – a guide to airless spraying" if you have not already got it, this is available on Amazon.

As part of the course you will be added to the "PTA support" group. This is a private Facebook group for people who have done the course. This means that it's a serious group and you have access to more information. PTA stands for "Paint Tech Academy" of course.

Finally, there is the free Facebook group "Spraying Makes Sense" this has over seven thousand members and is a very vibrant group. This is a great start for people who are thinking about spraying and are just dipping their toes in the water.

Spraying diary – with notes

Job and date	The name and date of the job
Area	Approx metres square
Paint	Brand, type and finish
Thinning	Did you thin it – how much?
Tip and pressure	Tip size, filter size and pressure you sprayed at.
Comments	This is just an example….

Spraying diary – blank

Job and date	
Area	
Paint	
Thinning	
Tip and pressure	
Comments	

Printed in Great Britain
by Amazon

76840573R00037